Armando Vinicio Paredes Peralta
Luis Fernando Arboleda Álvarez
Jhoanna Isabel Caiza Cuzco

STARCH, AS AN EDIBLE COATING IN FOOD PRESERVATION

Armando Vinicio Paredes Peralta
Luis Fernando Arboleda Álvarez
Jhoanna Isabel Caiza Cuzco

STARCH, AS AN EDIBLE COATING IN FOOD PRESERVATION

USE AND EFFECT ON FRUIT PRESERVATION

Imprint

Any brand names and product names mentioned in this book are subject to trademark, brand or patent protection and are trademarks or registered trademarks of their respective holders. The use of brand names, product names, common names, trade names, product descriptions etc. even without a particular marking in this work is in no way to be construed to mean that such names may be regarded as unrestricted in respect of trademark and brand protection legislation and could thus be used by anyone.

Cover image: www.ingimage.com

This book is a translation from the original published under ISBN 978-613-9-43547-0.

Publisher:
Sciencia Scripts
is a trademark of
Dodo Books Indian Ocean Ltd. and OmniScriptum S.R.L publishing group

120 High Road, East Finchley, London, N2 9ED, United Kingdom
Str. Armeneasca 28/1, office 1, Chisinau MD-2012, Republic of Moldova, Europe
Printed at: see last page
ISBN: 978-620-8-21400-5

"STARCH, USE AND EFFECT AS AN EDIBLE COATING IN FRUIT
PRESERVATION".

TABLE OF CONTENTS

TITLE

"STARCH, ITS USE AND EFFECT AS AN EDIBLE COATING IN FRUIT PRESERVATION".

SUMMARY

Post-harvest losses of fruits and vegetables caused by microorganisms are high due to the lack of technological resources and the absence of protection systems, which causes the low competitiveness of this value chain, directly affecting the economy of producers. For this reason, several researchers have focused on the search for new techniques that are environmentally friendly and extend the shelf life of products in the fruit and vegetable chain for a longer period of time. This paper focuses on compiling existing information on starch, its use and protective effect as an edible coating in fruit preservation. The information that composes the following document comes from books, magazines, standards and electronic theses, completing the search with the reading and tracking of bibliography referenced in the selected documents, in order to provide a good base and a global vision of the topic, which were prioritized according to the hierarchy of scientific evidence. Based on the results evidenced in different studies, it is deduced that starch is an interesting alternative for fruit preservation, because it acts as a protective barrier that prevents weight loss, preserves sensory characteristics for a longer time, and prolongs the shelf life of the fruit for a longer period of time. For this reason, it has come to be considered as a promising product to create edible coatings and films, as it is a highly available, easily biodegradable and environmentally friendly resource.

Key words: starch, fruits, edible coatings, postharvest, functional properties.

INTRODUCTION

Fruits are highly perishable agricultural products, due to the metabolic activity that continues even after being harvested, and at the same time to its high water content that facilitates the necessary living conditions for the development of fungi and bacteria, thus causing the rapid and extensive destruction of tissue throughout the anatomy of the product, reducing the quality and commercial value of the fruit. (Food and Agriculture Organization of the United Nations. 2004. p.8)

(Infoagro. 2019.p.1) states that: Globally, post-harvest losses of fruits and vegetables caused by microorganisms, are of the order of 5-25% in developed countries and 20-50% in developing countries. The difference lies in the fact that developed countries have greater availability of technological and economic resources to prevent losses. However, in developing countries, post-harvest losses are high due to the lack of technological resources and the absence of protection systems, which causes low competitiveness of this value chain, seriously limiting its improvement and directly affecting the economy of traders.

As mentioned above, several studies have focused on the research of different preservation techniques that help reduce food spoilage and perishability, referring to the use of edible coatings as an alternative for post-harvest preservation.

According to (Fernández, D. et al., 2015.p. 53), an edible coating (RC) can be defined as a thin, edible, continuous transparent matrix, which is formed around a food in order to preserve its quality, reduce the proliferation of microorganisms and serve as packaging. Most RC, are made from polysaccharides for their good adhesion properties to the fruits, which allows them to be more effective, being starch a carbohydrate widely used to coat various fruits and vegetables, because its application in fruits, does not produce risks of changes in flavor, such as color, besides being an easily extractable and low cost resource. (Ramos, M., 2018. pp.1-3).

Thus, in order to optimize not only productivity but also the maximum use of resources and taking into account the concern of new generations to consume healthy products free of preservatives and other chemical addictives, this research has set the following objectives: To compile existing information on starch, its use and its protective effect as an edible coating in fruit preservation. To select the most important theoretical foundations that allow knowing the benefits provided by coatings in fruit preservation. Identify through the use of the bibliographic review which starch is the most suitable for the elaboration of edible coatings.

CHAPTER I

1. THEORETICAL FRAME OF REFERENCE

1.1. Fruits

According to the Ecuadorian Institute of Normalization (INEN 1751, 1996, p. 4), on fresh fruits, it mentions that "A fruit is any edible organ of the plant, coming from the fruiting, destined for consumption in its natural state, whose cells are maintained in a state of turgidity and that present characteristics of commercial maturation".

1.1.1. *Post-harvest fruit problem*

After harvest, fresh fruits are susceptible to attack by saprophytic pathogens or parasites, due to their high water and nutrient content and because they have lost most of the intrinsic resistance that protects them during their development on the tree. Economic losses caused by postharvest diseases currently represent one of the main problems in world fruit growing (Acuña, L. et al., 2015. p. 1).

The most important pathogens that cause high losses of fruits and vegetables are usually bacteria and fungi, however, more frequently it is fungal species that cause pathological deterioration of fruits, leaves, stem and subway products (roots, tubers, corms, etc.) (Rubio, F., 2015. pp. 3-4).

Food loss and waste (FWL) is an increasingly relevant issue in Canada, the United States and Mexico, countries where nearly 170 million tons of food are lost and wasted annually, generating huge amounts of methane, a greenhouse gas 25 times more potent than carbon dioxide, in addition to other environmental and socioeconomic effects, such as inefficient use of natural resources, economic losses, loss of biodiversity and public health problems. (Commission for Environmental Cooperation, 2017.p.9)

1.1.2. *Diseases and deterioration*

All fruits, vegetables and roots are living plant parts that contain 65% to 95% water and whose life processes continue after harvesting. Their life after harvest depends on the rate at which they

4

consume their stored food reserves and the rate of water loss. When food and water reserves are depleted, the product dies and decays. Any factor that accelerates the process can cause the produce to become inedible before it reaches the consumer.

Fresh produce can become infected before or after harvest by diseases spread by air, soil and water. Some diseases can penetrate the intact skin of the produce, while others can only cause infection when a lesion is already present. This type of damage is probably the main cause of fresh produce losses (López, H.,2013. pp.31-32).

Postharvest spoilage by fungi and bacteria on fresh produce causes physical damage, increased water loss and respiration. Bacterial contamination is most commonly caused by contact with infected water or by contact with soil bacteria, while fungi proliferate by extension and cell division or by forming spores that are dispersed by air, water, animal vectors and insects. During storage, the product ages and the tissues are weakened by a gradual degradation of the cellular structure and integrity, being unable to withstand invasion, resulting in infection by pathogenic organisms (i.e., the infection is latent). (Rubio, F., 2015. pp. 2-3).

1.2. Starch.

According to (Melo, D. &et al.,2015. p.38), starch is a biopolymer of great importance composed of amylose and amylopectin, it is the major source of nutrition for animals and humans, it is an important raw material for industry, it is an abundant, renewable, biodegradable and low cost material, extracted from various natural sources such as tubers, cereals, legumes and immature fruits, which when hydrolyzed can generate products of greater commercial value.

1.2.1. *Main sources of starch*

Important conventional sources for obtaining flour and starch are cereals such as corn, wheat, rice and sorghum, and tubers such as potato and cassava; leaves and seeds of leguminous plants are also used. Currently, other non-conventional sources that present physicochemical, structural and functional characteristics of use in the industry are being explored, such as materials for biodegradable packaging, production of resistant starches, elaboration of food products and as substitutes for wheat and corn starches in bakery. (Montoya, J. et al.,2015. p.12).

1.2.2. *Starch composition*

Starch granules are composed of a mixture of two polymers: amylose and amylopectin. These polymers have the same basic structure, but differ in their length and degree of branching, which ultimately affects the physicochemical properties. Amylose is essentially a linear or sparsely branched polysaccharide with α-bonds (1-4) with a molecular weight of 105-106 and can have a degree of polymerization (DP) as high as 600. While amylopectin is a highly branched polymer with a molecular weight of 107-109 and α (1-4) (about 95%) and α (1-6) (about 5%) linkage and with an outstanding chain DP ~ 15, which is responsible for the crystallinity of the materials and this structure affects the physical and biological properties.(Shah, U., et al. 2015.p.2)

The lower the amylase percentage, the more stable and resistant to retrogradation (reorganization of amylose and amylopectin into a crystalline structure when the starch pastes are cooled). Table 1-1 details the % amylose of some starches. Cassava starch has a low tendency to retrograde and produces a very clear and stable gel (Cevallos, J., 2007.p. 49).

Table 1-1. Percentage of amylose in the most common starches.

% of Amylose	Starch Type
24 a 36	Corn
17 a 29	Wheat
8 a 37	Rice
18 a 23	Papa
16 a 19	Yucca

Source: (Cevallos, J., 2007.p. 50).

Carried out by: The authors, 2020.

1.2.3. *Starch extraction methods*

There are different methods of starch extraction from corn, wheat, cassava, potato or banana. The main and most general are: The dry method and the wet method, these methods are quite simple for the extraction of starch from cassava, plantain and a little simpler than those from cereals or corn.(Carrasco, L.; & Molocho, V.,2015. pp.3-4).

1.2.3.1. Dry method:

It basically consists of grinding the fruit after washing, obtaining flour from this process, for subsequent sieving to obtain starch. Taking into account operations that are carried out immediately, to develop the method and obtain a final product of quality and with characteristics that are desirable in starch (Carrasco, L.; & Molocho, V.,2015. p.4).

1.2.3.2. Wet method:

This method consists of crushing or size reduction of the guineo for and remove in liquid medium those components of the pulp that are relatively larger as fiber and protein, subsequently facilitating the removal of water by decantation at the end the settled material is washed to remove the last fractions other than starch and finally submit to the dry purified starch (Carrasco, L.; & Molocho, V.,2015. p.4).

1.2.4. *Physical differences of starches*

Cassava and potato starches swell rapidly at low temperature; their peak viscosity is also high, while the peak viscosity of corn and wheat starches are relatively low, because the granules are moderately swollen and require higher temperatures. Native starches are insoluble in water at temperatures below their gel point. Viscosity is measured in Brabender units, which reflects the paste consistency and properties under heating and cooling in a given time. Brabender viscosity curves are characteristic and different for each type of starch (Almidones de Sucre.,2015. p. 2).

Table 2-1 details the viscosity curves, the characteristics for each type of starch.

Table 2-1: Viscosity temperature of the most common starches.

Starch 95°C 20 min.	Gel temperature. °C 50°C 20 min.	Peak viscosity range
Yucca	54-66	800-1500
Papa	56-66	1000-2500
Corn	70-80	300-600
Wheat	75-85	200-500

Source: (Almidones de Sucre.,2015. p. 2).

Carried out by: The authors, 2020.

1.2.5. *Viscosity of starches*

Starch in the presence of water and with an adequate supply of energy undergoes a gelatinization process in which its crystalline structure breaks down from insoluble granules to a solution of its molecules, giving rise to viscous pastes. During this process, amylose molecules diffuse in water forming a gel, while amylopectin loses its crystalline order. This order-disorder transition that starch polymers undergo when subjected to heating is what impacts the processing, quality and stability of starch-based products (Salgado. R, et al.,2019. p.99).

The following table details the dough properties of the most common starches

Table 3-1: Paste or paste properties for some starches.

Property	Starch Starches			Factor
	Yucca	**Papa**	**Corn**	
Knowledge	Quick	Quick	Slow	Granule swelling speed.
Stability during cooking	Poor	Poor	Good	Fragility and solubility of the granules.
Peak viscosity	High	Very high	Moderate	Granule growth and solubility.
Gelation	Download	Download	Very high	Retrogradation of molecules
Consistency	Filamentosa	Filamentosa	Cut	Swollen granules, stiffness and retrogradation
Thickening	High	Very high	Moderate	Swollen granule size and attraction.
Shear strength	Poor	Poor	Moderate	Stiffness

Source: (Cevallos, J., 2007.p. 54).

Carried out by: The authors, 2020.

1.2.6. *Factors influencing the formation of gels.*

- Starch origin: According to the different types of grains, the longer the hydrogen bond, the stronger and more resistant the gel will be.
- Native starches are insoluble in water at temperatures below their gel point.
- The higher the starch concentration, the higher the viscosity achieved.

- Viscosity decreases with the presence of sucrose. Sucrose exerts a plasticizing effect by decreasing the gel strength. This occurs because sucrose interferes in the interactions with water, which has an affinity for sucrose and absorbs it.
- The starch structure becomes more integrated as it does not interact with water, so we will have to apply more temperature to break the starch paste.
- Fats also exert a plasticizing action because they form complexes that make the gel less resistant, less strong.(Contreras, M.et al.,2015. pp.4-5).

The following table shows certain factors affecting the starch paste

Table 4-1: Factors affecting the viscosity of a starch paste.

Concentration	Temperature (°C)	Time (minutes)	Revolutions (rpm)	Granule breakage	Viscosity (cps)
3	90	30	120	No	15
5	90	30	120	No	1080
5	90	30	1800	Dear	317
5	100	30	160	Dear	754
5	100	30	1800	Dear	90
10	90	30	120	No	11800
10	90	30	1000	Dear	7920
20	100	30	1000	Complete	18500
20	100	60	1000	Complete	9480

Source: (Cevallos, J., 2007.p. 55).

Realized by: The authors, 2020.

1.2.7. *Starch in the plastics industry.*

The development of materials based on organic polymers from biomass that are biodegradable, have focused on starch, which is an abundant material, economically competitive with petroleum. Among the main sources of starch for the industry we can mention: potato, wheat, rice, barley, oats and soybean (Ruiloba. I, et al.,2018. p.1). Modified starch has special properties which can be used for various purposes, such as bioplastics, since they are obtained with resources at very low costs and with simple production methods, they are cheaper than some synthetic polymers (Holguín, J.,2019. p. 34).

(Villada. H, et al.,2008. p.6), mentions that the transformation of granular starch is influenced by process conditions such as temperature and plasticizer content, in this case water and glycerol are the most commonly used. During the different thermoplasticization processes, given their action as a lubricant which facilitates the mobility of the polymeric chains. In addition, they retard the retrogradation of thermoplasticized products. According to (León, C., 2018.p.14), the application of starch as a bioplastic mainly requires the transformation of the semicrystalline structure into an amorphous homogeneous matrix, thus obtaining a manageable material that allows the molding of a film or coating. The disadvantages of native starches can be overcome by physicochemical transformations that allow obtaining a more attractive and versatile material for the industry.

Figure 1 shows a biodegradable plastic made from cassava starch.

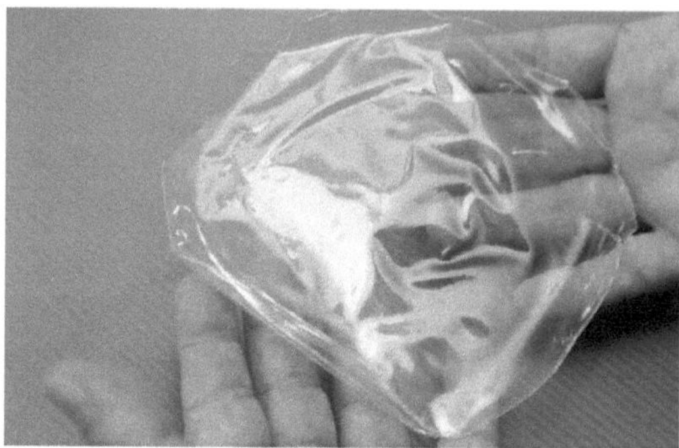

Figure 1-1: Biodegradable plastic from cassava starch

1.3. Edible coatings

They are solutions that adhere to the surface of the fruit, its components serve to improve appearance, increase brightness, preserve texture, softness and flavor. It is used as a means to add additives that control ripening, which extend shelf life and delay the physicochemical changes that occur in the fruit, the same that must be legal, harmless, sensory acceptable (Ruiz, D., 2015.p.45). For (Ancos, B. et al., 2015.p.10), these coatings are applied in liquid form by immersion or spraying, forming the film on the food.

This technology is environmentally friendly, and could replace, to some extent, plastic packaging with natural and biodegradable ones. Edible films and coatings improve the quality, safety and stability properties of coated foods. In addition, they modify the mechanical properties since they form a semi-permeable barrier to gases and vapors between the coated food and the surrounding atmosphere (Valencia, S.; & Torres, J., 2016. pp.162).

1.3.1. *Components of edible coatings*

PC and RC can be elaborated from a great variety of polysaccharides, proteins and lipids, alone or in combinations that take advantage of the benefits of each group, such formulations can include plasticizers and emulsifiers of different chemical nature that are used together in order to help improve the functional properties of the film or coating. They present benefits such as edibility, hardness, transparency, good barrier properties against oxygen and water vapor (Fernández, D., 2015. p. 54).

1.3.1.1. Lipids.

Lipids decrease the transmission of water vapor, preventing early dehydration of the fruit, and form a transparent solution that does not dull the characteristic color of the product. The mechanical barrier properties are very poor so it is necessary to mix with other substances (Ruiz, D., 2015.p.47).

1.3.1.2. Proteins

They are good materials for the formation of coatings as they show excellent mechanical and structural properties, but present a poor barrier capacity against moisture, which implies a decrease in respiration rate in fruits and vegetables, a situation that does not occur with lipids due to their hydrophobic properties, especially in those with high melting points, however, they present poor mechanical properties that must be counteracted with the use of additives (Fernandez, M. et al., 2017.p.136).

1.3.1.3. Polysaccharides

Polysaccharides are polymers containing hydroxyl groups of hydrophilic character, which is why they have high adhesion to food products. RC prepared with these components form coatings with a good barrier to gas exchange, but reduced protection to moisture loss. The components widely investigated for the formation of RC and PC are starch and its derivatives, chitosan, alginates, carrageenan, cellulose and its derivatives, pectins, and various gums. (Valencia, S.; & Torres, J., 2016. pp.163-164).

1.3.2. **Function of Edible Coatings**

According to (Solórzano, V., 2015.p. 30), one of the objectives for which edible coatings have been created, is to act in conjunction with synthetic packaging, in order to improve the sensory quality and extend the shelf life of food. Some of the functions that edible coatings perform, is the reduction of moisture loss, volatile components and gas exchange. (Solano, L. et al., 2018.p.32), mentions that the use of edible films and coatings on food, prevent the loss or gain of moisture that causes a modification in its texture and turgidity, delay chemical changes such as color, aroma and nutritional value, as they act as a barrier against gas exchange that influences the chemical and microbiological stability, in addition to avoiding mechanical damage by handling, also significantly reduces the loss of weight, water and gas exchange, as well as delaying aging and improving the sensory quality of these.

Coatings act as barriers to water loss and gas exchange, controlling the transfer of moisture, oxygen, lipids and flavor components (Figure 2), with an effect similar to that promoted by storage under controlled conditions or in modified atmospheres (Santiago, M., 2015. p.5).

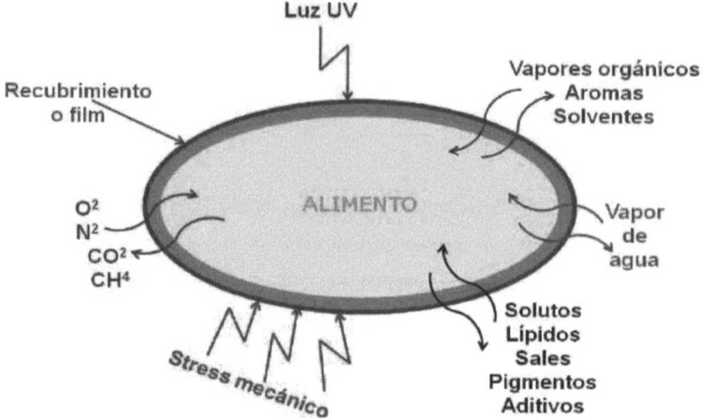

Figure 2-1: Transfers controlled by biodegradable films

Source: (Santiago, M.,2015. p.6).

1.3.3. *Additives used in edible coatings.*

(Falconi, J., 2016. p.45), states that an additive is several components that can be added to coatings to improve their physicochemical, mechanical, protective, sensory and nutritional properties to optimize their use.

Additives can be:

1.3.3.1. Plasticizers

Plasticizers are low volatility materials that are added to a polymer to increase its flexibility, elasticity and fluidity in the molten state. Among the most frequently used plasticizing agents are glycerol and sorbitol, which help improve mechanical properties, as well as water vapor permeability, thermal properties and sometimes color. The use of high concentrations of both plasticizers increases the percentage of elongation (Solano, L. et al., 2018. p. 38).

1.3.3.2. Emulsifiers

They prevent the fracture of the coating on the food, reduce water activity and the rate of moisture loss in the product. Emulsifiers must be food grade emulsifiers, which are usually vegetable or animal derived edible fatty acid esters and sources of polyols such as glycerol, propylene glycol, sorbitol and sucrose. The first requirement of an emulsifier food is that it should be non-toxic, non-carcinogenic and non-allergenic.(Falconi, J., 2016. pp. 45-46).

1.3.3.3. Antimicrobial agents.

They are substances that are formed by active elements that aim to prevent, destroy, impede the action of some harmful microorganism, the result of its antimicrobial function in the coating will depend on the degree of concentration. Within this group are citric acid, ascorbic acid, benzoic acid (Falconi, F., 2016. p. 48).

1.3.3.4. Bioactive Compounds.

Bioactive compounds are natural substances that aim to preserve the natural and nutritional properties of foods and enrich them naturally. Within this group are essential oils that possess antioxidant, antimicrobial properties, whose use has been noted to be used in research to improve post-harvest handling (Falconi, J. 2016. p. 48).

1.3.4. Quality parameters of edible films and coatings .

1.3.4.1. Water vapor permeability

Water vapor permeability (WVP) is one of the most important factors in the characterization of PC, since it depends largely on this to preserve the properties of food, the transfer of water vapor generally depends on the hydrophobic portion of the components of the film or edible coating, since through the movement of water vapor in the polymers, the transfer of moisture from the product to the environment is controlled, so it is sought to be as slow as possible (Solano, L. et al., 2018.p. 32).

Water transfer through the film (PVA) occurs in three stages: first, water vapor condenses and is deposited on the high water concentration side of the film surface; second, water molecules move through the film, driven by a concentration or activity gradient; and third, water evaporates from the other side of the film (Garcia, M.et al.,2018. p.39).

1.3.4.2. Mechanical properties

The mechanical properties of films and coatings depend mainly on the polymer used for their formation and the addition of structuring agents and additives. One of the most commonly used additives are plasticizers. The effect of these compounds on the mechanical properties of films produces a reduction in stiffness, which translates to lower maximum breaking strain and higher relative elongation (García, A., 2017. p.18). The mechanical properties usually evaluated to characterize PCs are tensile strength, which is the force required to break the film by stretching; elongation which implies the degree to which the film can stretch before breaking; for this, Young's modulus is used which measures the stiffness and compressibility of a structural material (Garcia, M. et al., 2018.p.39).

1.3.4.3. Surface properties

The surface properties of films and coatings refer mainly to the ability to cohere and adhere to the surface of food products. These two properties depend on the polymer's ability to form numerous, high-strength intermolecular bonds, so that the polymer chains are tightly bound together. Plasticizers, which reduce the level of cohesion, and surfactants, which improve the ability to adhere to the food surface, are used to improve these properties (Garcia, A., 2017.p.15).

1.3.4.4. Optical properties

The color and transparency of edible films and coatings is of great importance, since a consumer's purchasing decision is strongly affected by the external appearance of the food. In addition, the color and appearance of the product directly influences the acceptance or rejection of the product at the time of ingestion, and can even affect the sensory perception of the product. Therefore, the addition of active compounds in coatings is important so that they help to avoid affecting the color and appearance of the fruit to a great extent (García, A., 2017.p.15).

1.4. Use of starch to formulate CR

Starch is increasingly taken into account as one of the renewable sources with film-forming capabilities because they provide many advantages: easy availability, high extraction yield, biodegradability and incompatibility, which makes it a promising product for edible coatings and films (Shah, U. et al., 2015.p. 1). Starch granules contain two types of polymeric molecules: amylose and amylopectin. The former exhibits excellent properties to form strong, isotropic, odorless, tasteless and colorless films (Solano, L. & et al., 2018.p. 33). Starch-based coatings do not present a characteristic taste, odor or color, so when used on a food matrix they would not alter the sensory profile (León, C., 2015.p.14).

However, although starch is a cheap and abundant material, it presents a hydrophilic character which makes it highly sensitive to water and exhibits limited water vapor barrier properties, in addition the mechanical development of the coating is affected by retrogradation, since the double helices of amylose and amylopectin are cross-linked hardening the film (Zapador, M.; & Chiral, A., 2018. P. 2).

However, if the granules are transformed into a homogeneous amorphous matrix, the feasibility of their use is increased. This is done by gelatinization of the starch granules and the presence of a plasticizer. A higher presence of amylose molecules increases the opacity and thickness of the films, while a lower presence generates translucent and thinner films (Basiak, E. et al., 2017. p.1). According to (Castillo, C., 2015. p. 27). Cassava starch granules contain a small percentage of lipids compared to cereal starches (corn and rice). This favors cassava starch, since these lipids form a complex with amylose, which tends to repress the swelling and solubilization of starch granules.

1.4.1. *Starch most commonly used in edible coatings.*

Among the hydrocolloids used for the formulation of films are: polysaccharides, being modified cassava starch the most used, also alginates, pectins, chitosan and cellulose derivatives, as well as proteins such as casein, collagen and some whey proteins belong to this group. (Solórzano, V.,2015. p.12).

(Amaiz, S. et al., 2018. p. 140) mentions that cassava starch is the most used to elaborate edible coatings and films due to its high performance and easy adhesiveness to the food, besides providing an attractive gloss to the coating. Along with them, it is usual and indispensable to add a plasticizer such as glycerin to improve the flexibility of the edible coating.

(Pilataxi J., 2017. p. 9), indicates that cassava starch has an unusual viscosity, which allows it to form soft gels, odorless, transparent and relatively stable to retrogradation, this quality makes it interesting at industrial level. It is therefore widely used by the paper, textile and food industries. In its native state, starch granules have the property of increasing in size, and swelling very quickly at low temperatures (Castro, M. et al., 2017.p. 43). Cassava starch creates a modified atmosphere inside the fruit, reducing the rate of transpiration and delaying the senescence process, it also creates a barrier that limits the entry and exit of gases such as O2 and CO2, which delays the deterioration of the fruit by dehydration, helping to maintain the structural integrity of the food and retain volatile compounds (Simbaña, K.,2019. p. 4). By creating a barrier it decreases the presence of volatile substances such as ethylene, when dried it is able to form films with characteristics of resistance and transparency similar to those of cellulose, which makes them an option for the preservation of fruits and vegetables, in addition, if necessary it is easily removed and due to its non-toxicity it can be ingested without affecting the taste, aroma or appearance of food (Trujillo.C.,2014. p.20).

Cassava starch is the second starch source in the world after corn, but ahead of potato and wheat; it is mainly used unmodified, but it is also modified with different treatments to improve its properties of consistency, viscosity, stability to changes in pH, and temperature, gelification, dispersion and thus to be used in different industrial applications that require certain particular properties. The estimated extraction yield of cassava starch is approximately 57%, which indicates that it is effective (Zapata, D., 2019.p.16).

1.4.2. *Advantages and disadvantages of starch-based edible coatings*

Polysaccharides are the most widely used hydrocolloids in the food industry, since they are part of most of the formulations currently on the market, they form molecular networks cohesive by a high interaction between their molecules, this gives them good mechanical and gas barrier properties (O2 and CO2), so they retard respiration and aging of many fruits and vegetables (Fernandez, D., 2015. p. 54). Starches offer a very attractive low-cost base for new biodegradable polymers due to their low material cost and their ability to be processed with conventional plastic processing equipment. Commercialized starch-based products, which are not only fully biodegradable, but can also be used for animal feed or even edibles, are increasingly being taken into account as possible alternatives to solve problems such as oil shortages and the growing interest in alleviating the environmental burden due to the extensive use of petrochemical-derived polymers. (Tianyu J., et al.2019. p.8)

Among the renewable sources with film-forming capability, starch satisfies all the main aspects, such as easy availability, high extraction yield, biodegradability and biocompatibility, making it a promising product for edible coatings/films that exhibit odorless, tasteless, colorless, non-toxic, biologically absorbable, semi-permeable to carbon dioxide, moisture, oxygen, lipids and flavor components. The properties of the starch film are similar to the effect promoted by controlled or modified atmosphere storage and can be attributed to its chemical composition. Starch granules are composed of a mixture of two polymers: amylose and amylopectin.(Shah, U. et al., 2015. pp.1-2).

Although starch appears to be an ideal replacement for petroleum-based plastics, mainly due to its abundance, renewability, biodegradability and low cost, there are several disadvantages that make its application unfeasible. Poor mechanical behavior and high water vapor permeability (WVP) are the main drawbacks of starch-based materials, and are a consequence or starch structure.

On the other hand, due to its molecular structure, it has a relatively high glass transition temperature (Tg), and then brittle behavior at room temperature. This brittleness increases with time due to retrogradation. In addition, plasticizers such as glycerol have been shown to influence the crystallization kinetics of starch and thus the final mechanical properties of TPS. However, so far this biodegradable polymer cannot be used for broad applications due to some limitations. Compared to common thermoplastics, starch-based biodegradable products unfortunately reveal disadvantages that are mainly attributed to the highly hydrophilic character of starch. (Ribba L., et al.2017. pp.37-38).

1.4.3. *Preparation of starch-based coatings.*

Coatings made from cassava starch, have been applied in fruits such as: chonto tomato (Astudillo, J.; & Botina, K., 2017. p. 31) mango (Estrada, E. et al., 2015.p.182), apple (Pauta, D., 2018. pp. 6-7), pear (Castro et al, 2017.p.44) guava. (Amaiz, S. et al., 2018. p. 141) and elaborated as follows: the dilution of the starch is made in distilled water, mixture is brought to 82°C-95°C under constant agitation until reaching gelation or thermal coagulation (mechanism of elaboration of the hydrocolloid matrix of the coating), after the cooling period the additives such as glycerol and oils are added, the agitation is continuously homogenized for 5 minutes more, time in which a homogeneous and stable mixture is achieved. This methodology can vary if other starches are used, for example, Chinese potato starch applied in strawberries (Oñate, L., 2018.p.14) performed it as follows: the dilution of the starch is made and the mixture is carried 90°C for 5 minutes at 200 rpm, then the temperature is decreased to 70°C to add additives such as sorbitol and homogenized for 5 min more. The process will depend on the methodology chosen by the researcher.

In addition, (Fernández, M. et al., 2017.p.138) in his research work with the topic current status of the use of edible coatings on fruits and vegetables mentions that: 'there are coatings with mixtures of several components as is the case of corn starch with chitosan and sunflower essential oil for coatings on citrus fruits in which the preparation of film-forming solutions is done in stages where first a solution of starch with glycerol is made, then a solution of chitosan with acetic acid, then a mixed solution from the previous formulations of gelatinized starch with added glycerol and chito-sano at 12.500 rpm for 3 minutes, and finally a mixed solution with incorporation of a lipid phase by the addition of sunflower oil and Tween 80, the dispersion was carried out at 12,500 rpm.

1.4.4. *Coating application methods:*

The ways of applying coatings have been in continuous evolution so there are different methods, according to (Valencia, S.; & Torres, J., 2016.p. 165) there are certain aspects that must be taken into account before being applied: "be evenly distributed, dry quickly and be easy to remove and clean the equipment used for its formulation, they should not ferment, coagulate, generate undesirable odors or separate into phases".

(Ruiz, D., 2015.p.45), points out that edible coatings can be applied in ways such as: dipping and spraying.

And he mentions that the spraying method, "is carried out on foods with smooth surfaces, which must be washed and dried beforehand. This method is based on applying the pressurized solution, which allows obtaining thinner and more uniform coatings".

While (Fernández, M. et al., 2017.p. 138) states that: "one of the most widely used methods is immersion because it results in a uniform coating on irregularly shaped foods, for this the fruit should be washed and dried previously, then immersed in the coating formulation, the excess material should be allowed to drain and proceed to dry".

1.4.5. *Procedure for the application of starch-based coatings on fruits.*

The application of an edible coating on mango based on cassava starch (Estrada, E.et al., 2015.p.182) was performed by immersion method, immersing the fruits in the obtained solution for a period of 2 min, after the application they were left to dry for 120 min, under ambient laboratory conditions. For strawberry (Oñate, L., 2018.p.14), using Chinese potato starch coating, performed the immersion of the strawberry in the coating for 5 or10 min. Then, the strawberry was placed in a dryer for 45 min. Similarly (Pinzón, O.; & García, M., 2018.p.3), using banana starch on strawberry applied the immersion method, for 3 min, to then be dried in a hot air convection oven for 20-30 min. (Amaiz, S. et al., 2018. p. 141), using cassava starch in guava subjected the fruits to immersion in their respective edible coatings, for 5 min in the formulations.

It can be seen from the bibliographic review that starch coatings were mostly applied by the immersion method.

1.4.6. *Positive effect of the use of additives in starch CR*

Starch presents numerous benefits to be used as a base for coatings, however, (Pauta, D., 2018. pp. 4) mentions that this polysaccharide shows some limitations as a coating in food so the improvement of its characteristics is an important factor. To improve the mechanical properties of starch it is common to subject it to modifications, both physical and chemical, among the improvements that should be considered, is to use specific additives in the appropriate amounts, with the objective of improving its properties. For example, if chitosan is added to the biopolymer obtained from starch, it would increase its mechanical and barrier properties, as well as prevent the generation of fungi and bacteria on the surface of the material, due to its antifungal and antimicrobial properties, thus prolonging the shelf life of food (Alarcón, H.; & Arroyo, E., 2016.pp. 316-317). The following is a brief mention of some research that used additives and their effect.

(Basiak, E. et al., 2016. p.1) conducted a study in which they added rapeseed oil to starch-based biodegradable films, and obtained composite films that were more opalescent and glossier than starch films without fat. In addition, the addition of the oil significantly reduced water vapor and permeability. (Basiak, E. et al., 2018. p.1) indicated that water and glycerol content significantly influence the physical and functional properties of starch-based coatings and films. Furthermore, they concluded that glycerol content can strongly affect the functional properties of coatings, however, it does not play a significant role in color or mechanical properties. Thus (Song, X. et al., 2018.p.1) Evaluated the effect of essential oil and surfactant on the physical and antimicrobial properties of corn and wheat starch films, the authors reported that the incorporation of lemon essential oil caused a decrease in water content, transparency, whiteness index (WI), water vapor permeability (WVP), solubility and tensile strength properties. Films with LO, especially at higher concentrations, were more effective against all tested bacteria than control films.

1.4.7. *Recent studies highlighting the benefits of the use of starch in CR*

(Pinzón, O.; & García, M., 2018.p.5), evaluated an edible coating of 4 % guava banana starch on the quality of strawberries stored under refrigerated conditions, presenting favorable physicochemical characteristics (texture and weight loss) and extending the shelf life of the fruit by 5 days more in relation to the control sample. For their part (Astudillo, J.; & Botina, K., 2017. p. 60), made an edible coating of corn and cassava starch (3% and 4%) and evaluated its effect on the ripening of cherry tomatoes, the best result was the 4% cassava starch-based coating, which presented a superior shelf life of 15 days compared to the control treatment, improving the physical properties of the vegetable, a delay in weight loss, loss of firmness, had less respiratory activity due to its low O2 consumption and low CO2 release.

(Rocha, A. et al., 2017.p. 5), studied the ability of 2 coatings based on cassava starch and pectin to extend the shelf life of "Paluma" guavas. Fruits coated with pectin showed less weight loss compared to those coated with cassava starch. However, the cassava starch coating kept the fruits brighter compared to pectin-coated fruits. In a refrigerated, ready-to-eat ascorbic acid-enriched pumpkin product, edible coatings were applied to improve its stability (Genevois, C., et al. 2015. p.1). The edible coatings used were prepared with k-carrageenan or tapioca starch, with the addition of glycerol, potassium sorbate and iron. The results of these studies showed that the starch-based coatings significantly reduced the degradation of ascorbic acid in the coated product, while the control sample containing iron and ascorbic acid showed browning. The authors reported that the product obtained presented good color and texture characteristics, and was safe from the microbiological point of view.

CHAPTER II

2. METHODOLOGY

2.1. Methods for systematization of information

The present study is of a descriptive theoretical type. The methodological route followed basically comprised four moments: search, organization, systematization and analysis of electronic documents, without language restriction, related to the topic of starch-based edible coatings, of which 90% of the information belongs to the last 5 years and 10% corresponds to previous years.

In order to meet the proposed objectives, the research focused on a selective bibliographic review and a deep critical analysis of the data obtained related to the parameters of the study. To locate the documents, several documentary sources were used through the internet with the help of the "google academic" search engine, using the databases of journals such as: Revista Ciencias Técnicas Agropecuarias, Innovative Food Science and Emerging, Cogent Food & Agriculture, Scielo, Dianelt, International Journal of Biological Macromolecules, Revista Iberoamericana de Tecnología Postcosecha, among others. Much of the qualitative and quantitative information that makes up the following research comes from books, journals, technical reports, standards, theses, all electronic documents, and the search was completed with the reading and tracking of bibliography referenced in the selected documents, in order to provide a good basis and an overview of the topic, which were prioritized according to the hierarchy of scientific evidence.

As search criteria, the following descriptors were included: "starch", "edible coatings", "edible coatings", "starch", "effect of coatings". These keywords were combined in various ways at the time of exploration, with the aim of broadening the search criteria. The records obtained ranged from 30 to 40 records after combining the different keywords.

The information we worked with came from various sources, both primary and secondary. When searching for documents in each of the databases, several articles and documents were pre-selected from which those documents that were found to be most suitable were chosen in accordance with the inclusion and exclusion criteria. It is worth mentioning that those documents that do not comply with the adequate information will not be taken into consideration for the analysis.

21

2.1.1. *Selection criteria.*

For the analysis of the documents, some selection criteria were established, which were useful for the collection of information used during the research process, so the following parameters were proposed:

- Information with a high level of validity, i.e. in recognized and more "academically" valued formats such as: books, journals, conference proceedings, technical reports, standards, theses and the Internet.
- Updated content for the last 5 years.
- Analysis of the title, summary and results.
  With the purpose of knowing if the information they generate is useful and relevant and applicable to our subject of study.
- Accessibility to information.
- Quality information, that is, information that provides a good understanding of the text.
- Documents that are related to the stated objectives, that generate usefulness, availability and informative interest.
- Material, which supports the content of the different parts of the research.

In this document, tables, images and charts will be used to systematize the information.

3. RESEARCH RESULTS AND DISCUSSION.

3.1. Starch Use In Coatings

The potential use of starch as an edible coating material has been widely recognized for its good mechanical strength, low water permeability and because it acts as a barrier against gases, generating isotropic films, odorless, tasteless, colorless, non-toxic and biologically degradable, also improving gloss and opacity, decreasing shrinkage and increasing the stability of the freeze/thaw cycle (Solis, D. et al.,2015. p. 36).

(Sanchez, I., et al.2016. p. 1) mentions that the result obtained from different starch coatings is not only going to depend on the type of fruit or vegetable, to which it is applied, but also on the coating composition.

A description of the amylose content of starches commonly used for the production of edible coatings and films can be seen in the following table.

Table 5-3: Amylose content of the most common starches.

Variables	Alvis. et al., (2008).	FAO (1997)	Ibarra et alt. (2010) (Average value)	Cevallos, J., (2007) (Average value)	Hernandez, M, et al.,(2008)	Average
Yucca	14 %	-	28 %	17 %	17%	19%
Papa	24 %	23 %	23 %	21 %	21%	22.4%
Wheat	-	26 %	74 %	23 %	-	41%
Corn	-	28 %	28 %	30 %	28,3%	28.57%

Realized by: The authors, 2020.

As can be observed in Table 5, the amylose content in cassava starch shows a lower content with respect to the others; in some investigations it was reported that the amylose content, for native cassava starches, varied between 14 to 19 %. Therefore, the differences found in amylose content depend to a great extent on the source from which the starches are obtained.

The amylose content and the length and location of the branches in amylopectin are the main determinants of starch functional properties, such as water absorption, gelatinization and sticking, retrogradation and susceptibility to enzymatic attack (Ribba, L. et al., 2017. p.38). Starch origin influences optical properties and thickness: with more amylose, films are opalescent and thicker; with less, they are transparent and thinner.

The following shows the influence of amylopectin content in edible coatings, based on the starches represented in the table and its effect on papaya; (Achipiz, S., et al.2013. p. 98), evaluated a coating with 4% potato starch in guava, where it showed an acceleration in ripening and loss in quality, of the control sample, observing the effect on guava.. p. 98), evaluated a coating with 4% potato starch on guava, where he evidenced the acceleration in ripening and loss in quality of the control sample, observing that the coated fruit was the most efficient, increasing in 10 days the shelf life and avoiding weight loss; (Baldez, R.2016.p.59), using corn starch (3%), presented a loss of quality in the uncoated fruits from the third day while the fruits with corn-based coating was maintained during the 15 days with respect to the uncoated treatment. (Amaiz, S. et al., 2018. p. 137) reported an optimal texture during the 24 days of storage in fruits coated with cassava starch at (7%) while in uncoated fruits the texture was optimal until day 12, this parameter is important because it presents the loss of turgor due to the effect of fruit senescence.

According to the results reported, the corn coating exhibited a better barrier capacity and prolonged the shelf life of the papaya for a longer period of time, as well as the highest amount of amylose after wheat. This is justified according to what is mentioned by (Zapador, A.; & Chiralt, A., 2018.p. 7) who, in his research, states that the amylose-amylopectin ratio affects the functionality in starch coatings, due to the different structure of the films, starches high in amylose, exhibit better packed domains, better preserve weight losses and firmness for longer periods than medium amylose starch coatings.

It should be noted that each type of coating has its own properties defined according to the composition, which is why, according to several studies, after verifying the effectiveness they generate, starches are considered excellent sources for coatings, especially cassava starch.

(Andrade, J. et al.2014. p. 2), points out that cassava starch has been well received because it has a good appearance, is not sticky, is a resource of high availability in various parts of the world, is shiny and transparent, improves the visual appearance of the fruit and can be removed with water, which represents a potential alternative to be used in the preservation of fruits and vegetables.

3.2. Benefits of starch-based coatings

Among the benefits provided by edible coatings, here are the most important and most studied in starch-based CRs.

3.2.1. *Decreased weight loss*

The use of edible coatings (RC) has become an effective and environmentally friendly alternative method to extend their shelf life and protect them from harmful environmental effects, creating semi-permeable barriers to gases and water vapor, reducing respiration and weight loss and maintaining the firmness of the fresh product while providing gloss to the coating. Improving the stability, quality and safety of food, thus promoting the functionality of performance coatings beyond their barrier properties. (Zapador, A.; & Chiralt, A., 2018.p.2)

Table 6-3: Influence of starch CR on fruit weight loss.

Matrix	Additives	Fruit	Weight loss (%)		Difference in weight loss	Reference
			SR Fruit	Fruit CR		
Cassava starch (4%)	Citric acid Glycerin Cinnamon essential oil	Tomato 22 days	14.81%	8%	6.81	Barco, P et al. (2011)
Banana starch 4%	Glycerol Chitosan	Strawberry 8 days	24%	17%	7%	Pinzón, O.; & García, M., (2018).
Papa china (2%)	Sorbitol Potassium sorbate Citric acid	Strawberry 16 days	18,46%	12,67%	5.79%	Oñate, L (2018).
Modified cassava starch	Plasticizers sorbitol	Strawberry 8 days	53.44%	32.96%	20.48%	Franco, M., et al. (2016).
Yucca 4%	Beeswax glycerin and carboxymethylcellulose	Chontaduro 8 days	22%	11%	11%	Tosne, L. (2014)

Realized by: The authors, 2020.

Table 5 shows the weight loss of different fruits, with a greater loss in the control samples, as opposed to the treated fruits, which retained their weight for more days.

Of all the treated fruits, it is distinguished that the coating that best acted as a barrier preventing weight loss was the modified cassava starch applied on strawberries with a difference in weight loss of 20.48% of the fruit with coatings in relation to the control sample, this is because for the preparation of this coating modified starch was applied, which considerably improved the physical characteristics of the coating, preventing moisture loss. This is how other researchers stress the importance of using this type of starch in coatings (García, M.et al.,2018. p.35), mentions that modified starch has been used to stabilize functional properties, showing a homogeneous surface and a decrease in water vapor permeability, compared to the control.

(García,A.et al, 2017.p.6), indicates that weight loss in fruit is due to the possible exchange of gases during the process of respiration and transpiration that decrease the water content that, in a general way in fruits. Thus, (Rocha, A. et al., 2017.p. 3), mentions that fruits coated with polysaccharide-based films tend to slow mass loss because the gel applied on the fruit loses moisture before the coated food dries out.

This is in agreement with what has been reported by other researchers, who found that the application of edible coatings on fruits and vegetables inhibits weight loss, preventing texture changes and shrinkage of the surface, thus avoiding a negative effect on the shelf life of climacteric fruits and vegetables. This proves the efficiency of the coatings and evidences the potential use of starch in the elaboration of edible coatings.

3.2.2. *Prolongation of the useful life time.*

According to (León, E. 2015.p.27), shelf life is the time during which the food retains all its qualities. The end of a food's shelf life depends not only on maintaining minimum levels of contamination, but also on preserving its physicochemical (homogeneity, stability, structure) and organoleptic (texture, flavor, aroma) qualities.

Table 7-3: Starch coatings and their influence on the shelf life of fruits.

Matrix	Additives	Fruit	Shelf life (days)		Increased useful life (days)	Reference
			SR Fruit	Fruit CR		
Cassava starch 5%	Glycerol Anti-blighting agents Ascorbic acid Others	Plantain	20	32	12	Márquez, C., et al (2015)
Yucca 4%	Beeswax Glycerin others	Chontaduro	12	16	4	Tosne, L. (2014)
Yucca	Laurel wax, Olive oil, glycerol others	Tree tomato	12	17	5	Andrade, J. et al. (2014)

| Yucca 7% | Glycerin | Guava | 12 | 24 | 12 | Amaiz, S. et al., (2018). |
| Yucca 2.5% | Glycerol Chitosan glacial acetic acid | Guava | 4 | 14 | 10 | Ferreira, N., et al (2018) |

Carried out by: The authors, 2020.

As can be seen in Table 6, the edible coating based on cassava starch provided longer shelf life to the coated fruits, such is the case of banana and guava, which extended the shelf life for 12 days more than the control sample, their attributes such as color, aroma, weight loss were maintained during the 24 days.

This parameter is greatly influenced by the amount of starch used in the preparation of the coatings. In both fruits, a higher concentration of this biopolymer was applied with respect to the other formulations, so that a higher amount of starch creates thicker coatings that help maintain firmness and prevent easy loss of moisture content, thus decreasing energy metabolism, favoring the maintenance of firmness for a longer period of time, delaying the progress of fruit ripening and maintaining the sensory characteristics for more days.

This can be explained by referring to the concept (Cusme, K.; & Gómez, A.2016.p.25), who, in his study, mentions that, by increasing the starch concentration, the adherence and flexibility of the coating on the fruit surface is improved. However, starch concentrations of 2% caused apparently dehydrated and opaque fruits, in addition to being brittle and fibrous coatings. In turn (Ferreira, N., et al. 2018.p. 282) indicates that as the concentration of cassava starch in the suspension increased, the values presented a lower loss, due to the reduction of water loss from the fruit, caused by the increase in the thickness of the coating.

Compared to other coatings, the starch coating presents better characteristics and prolongs for more days the storage time of fruits. (Jimenes, A.2016. p.15.) studied a coating based on aloe vera in guava, reported an increase of 8 days more for the shelf life, compared to the sample. While (Amaiz, S. et al., 2018) applied a 7% cassava starch coating on the same fruit, being the most effective, because it increased the shelf life 12 days more than the control sample.

Thus, it can be said that starch coatings are a convenient alternative that allows maintaining the physical, chemical and sensory attributes of agricultural products, extending their shelf life and reducing post-harvest losses. Relating to what was mentioned (Toalombo, F.2014.p. 102), where it indicates that coatings reduce oxygen content, slowing respiration and allowing anaerobiosis to occur and increasing carbon dioxide which can effectively inhibit microbial growth and can extend shelf life

3.2.3. *Color preservation in the fruit ripening process.*

According to the literature, starch coatings form a barrier which decreases the rate of respiration slowing metabolic changes, delaying the disintegration of chlorophyll and decreasing the concentration of carotenes that give the yellow color to the fruit in this case, together there is evidence of a delay in the production of ethylene which accelerates these biochemical processes, which stimulate the genes that are responsible for the synthesis of certain enzymes, which degrade pigments that cause the color change in the fruit (Simbaña, K.,2019.p. 34).

One of the main characteristics to indicate the ripening process of fruits in the postharvest phase is the skin color. The loss of the green color of the peel is due to a rupture in the molecular structure chlorophyll, involving the enzyme chlorophyllase (Ferreira, N., et al. 2018.p.283).

Table 83: Influence of starch-based CRs on fruit coloration.

Matrix	Additives	Fruit	Temperature of alm.	Coloring		Reference
				SR Fruit	Fruit CR	
Yucca 6%	-	Guava "Paluma".	3°C	Yellow	Green Dark	Rocha, A. et al., 2017.
Yucca 2.5%	Glycerol Chitosan Glacial acetic acid	Guava	22 ± 2 ° C	Yellow	Green	Ferreira, N., et al (2018)
Yucca 4%	Glycerin Thyme essential oil	Bell pepper	25 ° C	Reddish green	Green	Bolaños, D.,(2014).
Banana and banana starch Avocado	Vinegar Glycerin Ac. Asc. Citric Acid Potassium sorbate	Papaya	7 °C	Green Of course	Green dark	Chapuel, A.; & Reyes, X,. (2019)
Yucca 15%	Glycerol	Kidney Tomato	18°C	Red	Pintón (pink)	Pilataxi J., (2017).

Carried out by: The authors, 2020.

Table 6 shows that edible coatings have a significant effect on the fruit, preventing the skin from changing color, unlike uncoated fruits, which show color changes.

Several studies report positive effects of these coatings. (Rocha, A. et al., 2017.p. 5), observed that guavas coated with cassava starch at concentrations of (4 and 6%) kept the peel color greener than the control (0%) promoting the delay of peel yellowing compared to the control. (Ferreira, N., et al. 2018.p.283) reported a similar result, of cassava starch coatings, which kept guava green during the 12 days of storage, while the skin color of the control group changed from green to yellow in only 4 d. (Bolaños, D.2014.p.798), indicates that the decrease of green color in peppers was notorious, starting from a green color percentage between 56.1 and 57.55 and ending between 11.17 % and 11.56 % on day 17. (Chapuel, A.; & Reyes, X. 2019.p.118) using avocado and banana seed starches for papaya formalized that fruits under refrigeration varied their dark color to lighter green color after 18 days. (Pilataxi J., 2017. p. 44) elaborated a 15% cassava starch coating at 22 days, evidenced less alteration in terms of its coloration, preserving them in maturity stage 4 (Medium Pintón) at 3°C, presenting a good commercial maturity for tomatoes, being within the quality parameters established by the (NTE INEN 2832. 2013. p.6). While the control fruits presented more pronounced changes in coloration, reaching the maximum degree of maturity at room temperature 18°C.

From the results evidenced in different studies, it can be deduced that edible starch coatings are an alternative to prevent postharvest loss of fruits, because they act as a good protective barrier that prevents the loss of water, volatile substances, slows the loss of their sensory characteristics, and prolongs the shelf life of the fruits for a longer period of time. Thus, starch is considered as a good material for edible coatings, being within the definition established by the standard (INEN 751:96. 2012. p. 3) where it mentions that a coating is one that protects the surface of the fruit with substances such as oils, vegetable waxes and other products in order to reduce wilting, wrinkling and improve the appearance.

CONCLUSIONS

According to the review of several bibliographic researches focused on the study of edible coatings, the following conclusions can be drawn:

Starch has a high potential for use in the elaboration of edible coatings according to the extensive number of studies found, which highlight the effectiveness of this polysaccharide, in addition to emphasizing its availability and biocompatibility as a raw material, which leads to consider it as an excellent alternative for the protection and preservation of fruits and vegetables.

Starch-based edible coatings have shown a growing popularity in recent years, due to the multiple benefits they generate, their use is a future alternative for the preservation of fruits and vegetables as it prevents water loss, helps to preserve the nutritional content, maintain firmness, and produces a delay in ripening, has barrier properties that prevent the flow of gases, showing a lower degree of deterioration in fruits, extending the shelf life, providing added value and improving product quality for a prolonged period of time.

Several studies use starch as a base to elaborate edible coatings, extracted from different sources such as; corn, potato, green, and yucca, corn starch according to its chemical composition is the most suitable, to create coatings, however, it is very little used because its cultivation is destined to food production, and very little to the research field. While cassava starch is the one that stands out the most to create these products, several authors report that cassava has attributes that make it have great acceptance in relation to the others because it is a resource of high availability in various parts of the world, has a high yield as raw material (it produces a much greater volume of corn starch), has easy adhesiveness on the product, and is also known for having properties that favor the formation of coatings, among them the ability to gel and ease of molding.

RECOMMENDATIONS

31

Polysaccharides, due to their composition, offer moderate oxygen permeability properties and good barrier properties; however, they present certain limitations in their wettability and mechanical properties. Therefore, it is recommended to search for new techniques to improve the functional characteristics of the mixture used to make CR, with the purpose of increasing the range of its applications in the industrial sector and to elaborate a commercially viable product.

It is recommended to carry out research that focuses on other cereals and tubers, in addition to cassava, as possible sources of starch to obtain CR, in order to exploit their use and value the properties of forgotten species such as sweet potato, frutipan, among others.

Edible coatings use certain antioxidants, antibacterial agents and functional ingredients in order to maintain the sensory and nutritional characteristics of fruits, so it is important to conduct research to determine the possible positive effects that these coatings can have on the consumer's health in the long term.

BIBLIOGRAPHY .

ACHIPIZ, SANDRA. et al. Effect of starch-based coating on guava (psidium guajava) ripening. Bíotecnología en el Sector Agropecuario y Agroindustrial [On line]. 2013, (Colombia). p. 98. Available at: http://www.scielo.org.co/pdf/bsaa/v11nspe/v11nespa11.pdf.

ACUÑA, Luis, et al. Manual de poscosecha de frutas.INTA.ediciones [online], (2019), (Argentina), p. 1. [Accessed: 07 May 2020]. ISBN 978-987-8333-12-0 (digital). Retrieved from: https://www.researchgate.net/publication/336899829_Manual_de_poscosecha_de_frutas

ALARCÓN, H & ARROYO, E. Evaluation of chemical and mechanical properties of biopolymers from modified potato starch. Rev Soc Quím [Online], (2016), (Peru), 82(3), pp. 316-317. [Accessed: 07 May 2020]. Retrieved from: http://www.scielo.org.pe/pdf/rsqp/v82n3/a07v82n3.pdf

SUCRE STARCHES. Cassava starch. [On line]. Colombia. 2015. p. 2. [Accessed: 29 July 2020]. Available at: http://www.almidonesdesucre.com.co/es/productos.html?limitstart=0

ALVIS Armando, et al. Physical-Chemical and Morphological Analysis of Yam, Cassava and Potato Starches and Determination of the Viscosity of Pastes. Información Tecnólogica [Online], 2008, (Colombia), 19 (1), p. 4. [Consulted:12 August 2020]. Available at: https://scielo.conicyt.cl/pdf/infotec/v19n1/art04.pdf

AMAIZ, S, et al. Effect of cassava starch-based edible coating on chemical and sensory parameters of guava hulls. Cumbres [Online], (2019), (Venezuela), 5(1), pp.137-140-141. [Accessed: 30 July 2020]. ISSN: 1665-0204. Available at: http://investigacion.utmachala.edu.ec/revistas/index.php/Cumbres/article/view/363

ANCOS, B, et al. Use of edible films/coatings in fresh-cut and pre-prepared convenience products. Revista Iberoamericana de Tecnología Postcosecha [Online], (2019), (Mexico): 16 (1), p. 10 [Accessed: 30 July 2020]. ISSN: 1390-9541. Available at: https://www.redalyc.org/pdf/813/81339864002.pdf

ANDRADE, Johana, et al. Development of an Edible Composite Coating for the Preservation of Tree Tomato (Cyphomandra betacea S.). La Serena [On line]. 2014, (Colombia), 25 (6). p. 1-2. ISSN 0718-0764. Available at:

https://scielo.conicyt.cl/scielo.php?script=sci_arttext&pid=S0718-07642014000600008

ASTUDILLO GÓMEZ, Jhon; & **BOTINA MACÍAS,** Karol. Elaboration of an edible coating based on corn and yucca starch for chonto tomato (lycopersicom esculentum Mill) [On line] (Degree Thesis) Universidad del Cauca, Facultad de Ciencias Agrarias, Programa de Ingeniería Agroindustrial. Popayán- Colombia. 2017. pp. 31-60-97. [Accessed: 2020-01-08]. Disponible en:

http://repositorio.unicauca.edu.co:8080/bitstream/handle/123456789/1722/ELABORACI%c3%93N%20DE%20UN%20RECUBRIMIENTO%20COMESTIBLE%20A%20BASE%20DE%20ALMID%c3%93N%20DE%20MA%c3%8dZ%20Y%20DE%20YUCA%20PARA%20TOMATE%20CHONTO%20%28Lycopersicom%20esculentum%20Mill%29.pdf?sequence=1&isAllowed=y

BALDEZ ALVES, Rafhaela. Revestimentos de amido, nanofibras de celulose e me-tabissulfito de sódio em goiabas (psidium guajaval.) Min-imamente processadas [On-line] (Degree Thesis) (Engineering). Federal University of LAVRAs. LAVRAS. 2016. p.59. [Accessed: 2020-08-19]. Available:

http://repositorio.ufla.br/jspui/bitstream/1/12087/2/DISSERTA%C3%87%C3%83O_Revestimentos%20de%20amido%2C%20nanofibras%20de%20celulose%20e%20metabissulfito%20de%20%20s%C3%B3dio%20em%20goiabas%20%28Psidium%20guajava%20L.%29%20minimamente%20processadas.pdf

BARCO HERNANDEZ, Paola Liceth, et al. Effect of a coating based on modified cassava starch on tomato ripening. Rev. Lasallian Research [On line]. 2011, (Colombia),8(2), p. 4. ISSN 1794-4449. Available in: http://www.scielo.org.co/pdf/rlsi/v8n2/v8n2a11.pdf

BASIAK, E. et al. Effect of Oil Lamination Between Plasticized Starch Layers on Film Properties. International Journal of Biological Macromolecules. [Online], (2016), (USA): 195 (1), p.1 [Accessed: 25 March 2020]. DOI: oi.org/10.1016. Available at: https://www.sciencedirect.com/science/article/abs/pii/S0308814615006445

BASIAK, E. et al. Effect of starch type on the physico-chemical properties of edible films. International Journal of Biological Macromolecules 98 [Online], (2017), (USA): 98 (4), p.1 [Accessed: 25 April 2020]. DOI: 10.1016/j.ijbiomac.2017.01.122. Available at: https://doi.org/10.1016%2Fj.ijbiomac.2017.01.122

BASIAK, E. et al. How Glycerol and Water Contents Affect the Structural and Functional Properties of Starch-Based Edible Films. Polymers (Basel) [Online], (2018), (USA): 10 (4), p.1 [Accessed: 25 March 2020]. DOI: 10.3390/polym10040412. Available at: https://www.ncbi.nlm.nih.gov/pmc/articles/PMC6415220/

BOLAÑOS Ordoñez, D. Effect of modified cassava starch and thyme oil coating applied to bell pepper (Capsicum annuum). Mexican Journal of Agricultural Sciences [On line]. 2014, (Colombia),5 (5), p. 798. Available at : http://www.scielo.org.mx/pdf/remexca/v5n5/v5n5a6.pdf

CARRASCO HUANCA, L. & MOLOCHO VÁSQUEZ. Starch extraction. Universidad Nacional Autónomo de Chota, Carrera Profesional de Ingeniera Agroindustrial. [Online], (2015), (Chota-Ecuador), pp.3-4 [Accessed: 08 August 2020]. Available at: https://es.calameo.com/read/005193087c8fe3b2314cf

CASTILLO SANTOS, Cynthia Yvory. Rheological and physicochemical characterization of pastes and gels obtained from the starch of three varieties of native potato (Solanumspp.) [On line] (Engineering) (Undergraduate Thesis) Universidad Nacional del Altiplano, Facultad de Ciencias Agrarias, Escuela Profesional de Ingeniería Agroindustrial. Puno -Peru. 2017. p. 27. [Accessed: 2020-07-13]. Available at: http://repositorio.unap.edu.pe/bitstream/handle/UNAP/10072/Castillo_Santos_Cynthia_Yvory.pdf?sequence=1&isAllowed=y

CASTRO GARCÍA Marlón et al. Edible coating of chitosan, cassava starch and cinnamon essential oil to preserve pear (Pyrus communis L. cv. "Bosc"). La técnica [Online], (2019),

(Ecuador), p.43-44. [Accessed: 05 August 2020]. SSN: 1390-6895. Available at: https://revistas.utm.edu.ec/index.php/latecnica/article/view/970/910.

CEVALLOS CEDEÑO, José Antonio. The production and export of cassava starch from the province of Manabí and its demand in the Colombian market in the period 2002-2006 [On line] (Master's Degree) (Undergraduate Thesis). Universidad Laica Eloy Alfaro de Manabí, Center for Graduate Studies, Research, Relations and International Cooperation, Cepirci. Manabí - Ecuador. 2007. p. 50-54-55 [Consultation: 2020-07-29]. Available at: https://repositorio.uleam.edu.ec/bitstream/123456789/1264/1/ULEAM-POSG-FCI-0021.pdf

CHAPUEL TARAPUEZ, Andrea Yesenia & REYES SUÁREZ Jetzy Xiomara. Obtaining a biodegradable film from avocado (persea americana mill) and banana (musa acuminata aaa) seed starches for papaya coating. [On line] (Degree thesis) (Engineering). University of Guayaquil, Faculty of Chemical Engineering, Chemical Engineering Career. Guayaquil- Ecuador. 2019. p. 97-118. [Accessed: 2020-08-03]. Available at: http://repositorio.ug.edu.ec/bitstream/redug/39933/1/401-1355%20-%20Obtenc%20pelicula%20biodegradable%20partir%20almidones%20semilla%20de%20aguacate.pdf

COMMISSION FOR ENVIRONMENTAL COOPERATION (CEC). Characterization and Management of Food Loss and Waste in North America, Synthesis Report, Commission for Environmental Cooperation, Montreal. Canada. 2017. p. 9 [Accessed: 14 July 2020]. ISBN: 978-2-89700-228-2 Available from: http://www3.cec.org/islandora/es/item/11772-characterization-and-management-food-loss-and-waste-in-north-america-es.pdf.

CONTRERAS ESTRADA, M, et al. Gelatinization and gelification of starches. [Research] [Online]. Universidad Nacional Del Callao, Peru: 2015. pp.4-5 [Accessed: 08 August 2020]. Available at: https://www.academia.edu/17812160/04_GELATINIZACION_Y_GELIFICACION_DE_ALMIDONES

CUSME RIVASANA, Karina Elizabeth & GÓMEZ SALVADOR, Ana Sofía. Percentages of starches with addition of natural plasticizers in the elaboration of a coating [On line] Report.

Manabí - Ecuador. 2019. p. 25. [Accessed: 2020-08-18]. Available at:
http://repositorio.espam.edu.ec/bitstream/42000/1062/1/TTMAI8.pdf

ESTRADA, E. et al. Effect of protective coatings on postharvest quality of mango (mangifera indical.). Rev. U.D.C.A. [Online], (2015), (Colombia), 18 (1), p.182 [Accessed: 25 April 2020]. ISSN: 181-188. Retrieved from: http://www.scielo.org.co/pdf/rudca/v18n1/v18n1a21.pdf

FALCONÍ NOVILLO, José Francisco. Use of edible coatings in the preservation of *Fragaria x Ananassa* (FRESA) [On line] (Degree Thesis) (Engineering) *(*Degree Thesis). Higher Polytechnic School of Chimborazo, Faculty of Sciences, Livestock Industry Engineering. Riobamba- Ecuador. 2016. p. 45-46-48. [Accessed: 2020-06-16]. Available at:
http://dspace.espoch.edu.ec/bitstream/123456789/6109/1/27T0331.pdf

FERNÁNDEZ VALDÉS, Daybelis, et al. Edible films and coatings: a favorable alternative in postharvest preservation of fruits and vegetables. Revista Ciencias Técnicas Agropecuarias, [Online], 2015,(Cuba), 24 (3), pp. 53-54. [Accessed: 28 July 2020]. ISSN 1010-2760. Available at:
https://www.researchgate.net/publication/317517584_Peliculas_y_recubrimientos_comestibles_una_alternativa_favorable_en_la_conservacion_poscosecha_de_frutas_y_hortalizas

FERNÁNDEZ, Marcela. et al. Current status of the use of edible coatings on fruits and vegetables. Biotechnology in the Agricultural Sector. [Online], 2017, (Colombia), 15 (2), pp. 136- 138. [Accessed: 28 July 2020]. SSN - 1692-3561. Available at:
http://scielo.sld.cu/scielo.php?script=sci_arttext&pid=S2071-00542015000300008

FERREIRA SOARES, N, et al. Antimicrobial edible coating in post-harvest conservation of guava1 [Online]. 2018, (Brazil), 281 (289), p. 282- 283. Available at:
https://www.researchgate.net/publication/329019781_Antimicrobial_edible_coating_in_post-harvest_conservation_of_guava

FRANCO, M, et alt. Effect of Plasticizer and Modified Starch on Biodegradable Films for Strawberry Protection. Insitute of Sciencie and technology [Online]. 2016, p. 1.

https://doi.org/10.1111/jfpp.13063. Available at: https://ifst.onlinelibrary.wiley.com/doi/abs/10.1111/jfpp.13063

GARCÍA Mónica. et al. Modern methods for the characterization of edible films and coatings. BioTecnología. [Online], (2018), (Mexico), 22(1), p.39. [Accessed: 04 August 2020]. Available at: https://smbb.mx/wp-content/uploads/2018/06/Garci%CC%81a-et-al-2018.pdf

GARCÍA ROLDÁN, Aitor. Comparison of the physical properties of edible films based on whey protein isolate or fish gelatin with sea fennel extracts incorporated [On line] (Degree Thesis) (Engineering) (Degree Thesis). Public University of Navarra, School of Agricultural Engineering. Pamplona, Spain. 2017.p.15 [Accessed: 2020-08-03]. Available at: https://academica-e.unavarra.es/bitstream/handle/2454/29023/TFG%20A.%20Garcia.pdf?sequence=1&isAllowed=n

GARCÍA, O & PINZÓN, M. Effect of guava plantain (musa paradisiaca l.) starch coatings on strawberry quality. Revistas alimentos hoy [Online], 2016, (Colombia), 25 (39), pp. 3-5-9. [Accessed: 8 March 2020]. ISSN 2027-291X. Available at: https://alimentoshoy.acta.org.co/index.php/hoy/article/view/407

GENEVOIS, Carolina. et al. Application of edible coatings to improve global quality of fortified pumpkin. Innovative Food Science and Emerging Technologies [Online], 2017, (Argentina), vol. 33, p. 1. [Accessed: 11 June 2020]. Doi.org/10.1016/j.ifset.2015.11.001. Available at: https://www.sciencedirect.com/science/article/pii/S1466856415002143

HERNANDEZ Marilyn. Physicochemical characterization of tuber starches grown in Yucatan, Mexico. Ciênc. Tecnol. Aliment., Campinas [Online], 2008, (Mexico), 28 (3), p.5. [Consultation:12 August 2020]. DOI 10.1016/j.foodhyd.2017.05.023. Available at: https://www.scielo.br/pdf/cta/v28n3/a31v28n3.pdf

HOLGUÍN CARDONA, Juan Sebastián. Obtaining a bioplastic from potato starch [On line] (Engineering) (Degree Thesis). Fundación Universidad de América, School of Engineering, Chemical Engineering Program. Bogotá-Colombia. 2019.p.34 [Accessed: 2020-08-03].

Available at: https://repository.uamerica.edu.co/bitstream/20.500.11839/7388/1/6132181-2019-1-IQ.pdf

IBARRA HERNÁNDEZ, E. *Tequila engineering.* [On line]. 1st Edition, Bogota-Colombia: INNOVA, 2010. [Accessed: 12 August 2020]. Available at: https://books.google.com.ec/books?id=SytHZWErOv8C&pg=PA70&dq=Content+of+Amylose +and+amylopectin+in+the+common+starch+m%C3%A1s+&hl=en&sa=X&ved=2ahUKEwjrjt Gb6JbrAhWirVkKHbxbBRoQ6AEwAXoECAEQAg#v=onepage&q=Content%20of%20Amyl ose%20and%20amylopectin%20in%20the%20starch%20m%20m%C3%A1s%20common%20s &f=false

INFOAGRO. Deterioration of fresh fruits and vegetables in the postharvest period [Online] 2019, [Accessed: 2020-05-05]. Available at:

https://www.infoagro.com/frutas/deterioro_poscosecha_frutas_hortalizas.htm

ECUADORIAN STANDARDIZATION INSTITUTE (INEN). Ecuadorian Technical Standard INEN 1751. *Fresh fruits definition and classification.* Quito - Ecuador. 1996. p.3- 4 [Consultation: July 27, 2020]. Available at:

https://archive.org/details/ec.nte.1751.1996/page/n9/mode/2up

ECUADORIAN STANDARDIZATION INSTITUTE (INEN). Ecuadorian Technical Standard INEN 2832. Standard for tomato (codex stan 293-2007, mod). Quito - Ecuador. 2013. p.6 [Accessed: 21 August 2020]. Available at:

https://www.normalizacion.gob.ec/buzon/normas/nte_inen_2832.pdf

JIMENES TRUJILLO, Angélica. Edible coating based on aloe vera (aloe barbadensis miller) for papaya (carica papaya) and guava (psidi um guajava) as convenience foods. FICAYA [Online]. 2016, (Ecuador), p. 15. Available at:

http://repositorio.utn.edu.ec/bitstream/123456789/6455/2/ARTICULO.pdf

LEÓN CHUMBIAUCA Etelvina Carmen. Determination of the shelf life of fruits immersed in two types of gels at ambient t0 in seasonal periods [On line] (Master's Degree Thesis) (Degree

Thesis). Universidad Nacional Del Callao, Vicerrectorado De Investigación, Facultad De Ingeniería Pesquera Y De Alimentos. Bella Vista-Callao. 2015. p.27. [Accessed: 2020-08-19]. Available at: http://repositorio.unac.edu.pe/bitstream/handle/UNAC/991/005.pdf?sequence=1&isAllowed=y

LEÓN VIRGÜEZ, Carolina. Edible starch-based coatings with potential application in fruit preservation [On line] (Engineering) (Degree Thesis). Universidad Nacional Abierta Y A Distancia, Specialization in Food Processes and Biomaterials. Bogotá-Colombia. 2018.p.14 [Accessed: 2020-08-03]. Available at: https://repository.unad.edu.co/bitstream/handle/10596/21299/52975967_.pdf?sequence=1&isAllowed=y

LÓPEZ LÓPEZ, Henry. Behavior of jalapeño chile (Capsicum annuum) fruits to the disinfection with different sanitizers in Postharvest [On line] (Degree Thesis) (Engineering). Universidad Autónoma Agraria Antonio Narro, Engineering in Food Science and Technology. Coahuila-Mexico. 2013. pp 31-32. [Accessed: 2020-06-16]. Available at: http://repositorio.uaaan.mx:8080/xmlui/bitstream/handle/123456789/526/62620s.pdf?sequence=1

MÁRQUEZ CARDOZO, C, et al. Effect of cassava starch coatings with ascorbic acid and N-acetylcysteine on the quality of harton plantain (Musa paradisiaca). Revista Facultad Nacional de Agronomía [On line]. 2015, (Colombia),68 (2), p. 6. ISSN 0304-2847. Available at : https://translate.google.com/translate?hl=es&sl=en&u=http://www.scielo.org.co/scielo.php%3F script%3Dsci_arttext%26pid%3DS0304-28472015000200010&prev=search&pto=aue

MELO SABOGAL, Diana. et al. Exploitation of plantain (musa paradisiaca spp) pulp and peel to obtain maltodextrin. Biotechnology in the Agricultural and Agroindustrial Sector [Online], 2015, (Colombia), 13 (2), p. 38. Available at: http://www.scielo.org.co/pdf/bsaa/v13n2/v13n2a09.pdf.

MONTOYA LÓPEZ, J. & et al. Characterization of flour and starch from Gros Michel banana fruits (Musa acuminata AAA). Acta Agronómica, [Online], 2015, (Colombia), 64 (1), p. 12.

[Accessed: 14 July 2020]. ISSN 0120-2812 Available at: https://www.redalyc.org/pdf/1699/169932884002.pdf.

OLIVEIRA, Rodrigo. Fabrican plástico biodegradable con almidón de yuca [Online].Colombia : El Espectador, 2019. [Accessed: 4 August 2020]. Available at: https://www.elespectador.com/noticias/ciencia/fabrican-plastico-biodegradable-con-almidon-de-yuca/

OÑATE ZÚÑIGA, Lizbeth Estefanía. Development of an edible coating for strawberry (Fragaria x ananassa Duchesne) based on starch from Chinese potato (Colocasia esculenta Schott) of the white variety [On line] (Engineering) (Degree Thesis). Technical University of Ambato, Faculty of Food Science and Engineering, Food Engineering Career. Ambato-Ecuador. 2018.p. 14 [Accessed: 2020-05-16]. Available at: https://repositorio.uta.edu.ec/bitstream/123456789/28391/1/AL%20685.pdf

UNITED NATIONS AGRICULTURE AND HEALTH ORGANIZATION (FAO). Carbohydrates in human nutrition. Joint FAO/WHO Expert Consultation Report Rome. [Online], 1997, (Rome), p.75. [Accessed: 12 August 2020]. ISBN 92- 5-304114-5. Available at: https://books.google.com.ec/books?id=FZ_ed5pkNdoC&pg=PA74&dq=Content+of+Amylose+and+amylopectin+in+the+common+starch+m%C3%A1s+&hl=en&sa=X&ved=2ahUKEwiG76Tt0pbrAhVQmVkKHW98BuwQuwUwAHoECAAQCQ#v=onepage&q=Content%20of%20Amylose%20and%20amylopectin%20in%20the%20starch%20m%20C3%A1s%20commons&f=false

UNITED NATIONS AGRICULTURE AND HEALTH ORGANIZATION (FAO). Fruit and vegetable preservation by combined technologies. Training manual. [Online], 2004, (Rome), p.8. [Accessed: 24 August 2020]. Available at : http://www.fao.org/3/a-y5771s.pdf

PAUTA LUNA, Diego. Edible coatings based on starch and gelano gum for the postharvest preservation of apple [On line] (Master's Degree Thesis). Universitat Politécnica De Valencia, Food Engineering. Valencia- Spain. 2017.pp. 4-6-7. [Accessed: 2020-05-16]. Disponible en: https://riunet.upv.es/bitstream/handle/10251/99194/PAUTA%20-%20RECUBRIMIENTOS%20COMESTIBLES%20A%20BASE%20DE%20ALMID%C3%93N%20Y%20GOMA%20DE%20GELANO%20PARA%20LA%20CONSERVACI%C3%93N%20POSTCO....pdf?sequence=1.

PILATAXI RAMÍREZ Jorge. Effect of coating with three cassava starch solutions on the preservation of kidney tomato fruit (Solanumlycopersicum, Mill) [On line] (Degree Thesis) (Engineering). Universidad Central Del Ecuador, Faculty of Agricultural Sciences, Agricultural Engineering. Quito- Ecuador. 2019.pp.9-44. [Accessed: 2020-08-08]. Available at: http://www.dspace.uce.edu.ec/bitstream/25000/19846/1/T-UCE-0004-CAG-158.pdf

RAMOS GARCÍA, M. et al. Modified starch: Properties and uses as edible coatings for the preservation of fresh fruits and vegetables. Revista Iberoamericana de Tecnología Postcosecha [Online], (2018), (Mexico), 19(1), p.3. [Accessed: 05 August 2020]. ISSN: 1665-0204. Disponible en https://www.redalyc.org/jatsRepo/813/81355612003/81355612003.pdf

RIBBA Laura. et al. Disadvantages and limitations of starch-based materials. Research gate [Online], (2017), (Argentina), p.37-38. [Accessed: 05 August 2020]. Doi org /10.1016/B978-0-12-809439-6.00003-0 Available from· https://www.sciencedirect.com/science/article/pii/B9780128094396000030

ROCHA Anny, et al. Conservation of "Paluma" guavas coated with cassava starch and pectin. DYNA. [Online]. 2017, (Colombia), 85 (204), p. 3-5. [Accessed: 3 July 2020]. ISSN 0012-7353. Available at: http://www.scielo.org.co/scielo.php?pid=S0012-73532018000100344&script=sci_abstract&tlng=pt

RUBIO ARAUJO, Favio Nolberto. Methods of post-harvest control of anthracnose in fruits [On line] (Degree Thesis) (Engineering). Universidad Nacional de Trujillo, Facultad de Ciencias Agropecuarias, Escuela Académico Profesional de Ingeniería Agroindustrial. Trujillo - Peru. 2015. pp. 2-3-4. [Accessed: 2020-06-16]. Available at: http://dspace.unitru.edu.pe/bitstream/handle/UNITRU/4302/RUBIO%20ARAUJO%20FABIO%20NOLBERTO.pdf?sequence=3&isAllowed=y

RUILOBA, Ivanova. et al. Elaboration of bioplastic from mango seed starch. Grupo Ciencia y Tecnología Innovadora de Alimentos [Online], 2018, (Panama), 4(1), p.1. [Accessed: 03 August 2020]. Disponible en: file:///D:/TEMPOR~1/1815-Texto%20del%20art%C3%ADculo-8742-2-10-20180710.pdf

RUIZ MEDINA, DOLORES. Design of a bioactive edible coating to be applied on strawberry (fragaria vesca) as a post-harvest process [On line] (Degree Thesis) (Engineering) (Degree Thesis). National Polytechnic School, Faculty of Chemical Engineering and Agroindustry. Quito- Ecuador.2015. p.45- 47. Available at:
https://bibdigital.epn.edu.ec/bitstream/15000/11181/1/CD-6412.pdf

SALGADO ORDOSGOITIA, R. et al. Analysis of gelatinization curves of native starches of three yam species:criollo (dioscorea alata), espino (dioscorea rotundata) and diamante 22. Scielo, [Online], 2019, (Colombia), 30(4), p.99. [Accessed: 30 July 2020]. Available at:
https://scielo.conicyt.cl/pdf/infotec/v30n4/0718-0764-infotec-30-04-00093.pdf

SÁNCHEZ, I, et al. Characterization and antimicrobial effect of starch-based edible coating suspensions. Food Hydrocolloids [Online], 2016, (USA), p. 1. [Accessed:14 August 2020]. Available from: https://www.sciencedirect.com/science/article/abs/pii/S0268005X15300795.

SANTIAGO SANTIAGO, Maricela. Elaboración y caracterización de películas biodegradables obtenidas con almidón nano estructurado [En línea] (Trabajo de Titulación) (Master), Universidad Veracruzana, Instituto De Ciencias Básicas. Veracruz- Mexico. pp.5-6. 2015. Available at:
https://cdigital.uv.mx/bitstream/handle/123456789/46809/SantiagoSantiagoMaricela.pdf?sequence=2&isAllowed=y

SHAH, Umar. et al. A review of the recent advances in starch as activeand nanocomposite packaging films. Cogent Food & Agriculture, [Online], 2015, (USA), 1(1), pp.1-2. [Accessed: 05 August 2020]. ISSN (Print) 2331-1932 (Online) Journal homepage: Available from:
https://www.tandfonline.com/doi/pdf/10.1080/23311932.2015.1115640

SIMBAÑA TIPÁN, Klever Fernando. Evaluation of the effect of coating with two cassava starch solutions on babaco (Vasconcellea x heilbornii. Heiborn) at two temperatures [On line] (Engineering) (Degree Thesis). Universidad Central Del Ecuador, Faculty of Agricultural Sciences, Agricultural Engineering. Quito- Ecuador. 2019.p.4-30-34. [Accessed: 2020-08-08]. Available at: http://www.dspace.uce.edu.ec/bitstream/25000/18335/1/T-UCE-0004-CAG-080.pdf

SOLANO DOBLADO, L. et al. Functionalized edible films and coatings. TIP Revista Especializada en Ciencias Químico-Biológicas, [Online], 2018, (Mexico), 21(1), p.32-33-38. [Accessed: 2 July 2020]. DOI 10.22201/fesz.23958723e.2018.0.153. Available at: http://tip.zaragoza.unam.mx/index.php/tip/article/view/153/166.

SOLÍS JIMENEZ, Diana et al. Effect of modified cassava starch coating on avocado hass. Rev. P+L [online], 2015, (Colombia), 10 (2), pp.32-36. ISSN 1909-0455.Available at: http://www.scielo.org.co/pdf/pml/v10n2/v10n2a04.pdf

SOLÓRZANO VILLACRÉS, Vicky. Study of the effect of an edible coating with sande (brosimum utile) latex on the shelf life of cassava (manihot sculenta), tree tomato (solanum betaceum) and potato (solanum phureja) [On line] (Degree Thesis) (Engineering) (Engineering). Escuela Superior Politécnica de Chimborazo, Faculty of Sciences, School of Biochemistry and Pharmacy. Riobamba- Ecuador. 2015.pp.12-30 [Accessed: 2020-06-16]. Available at: http://dspace.espoch.edu.ec/handle/123456789/4372

SONG Xiaoyong. et al. Effect of Essential Oil and Surfactant on the Physical and Antimicrobial Properties of Corn and Wheat Starch Films. International Journal of Biological Macromolecules. [Online]. 2018, (China), 107 (2), p.1 [Accessed: 8 July 2020]. DOI 0.1016/j.ijbiomac.2017.09.114. Available from: https://pubmed.ncbi.nlm.nih.gov/28970166/

TIANYU Jiang. et al. Starch-based biodegradable materials: Challenges and opportunities. Centre for Polymers from Renewable Resources. [Online], (2019), (China), 8(18), p.8. [Accessed: 05 August 2020.DOI 10.1016/j.aiepr.2019.11.003 Available at: https://www.sciencedirect.com/science/article/pii/S254250481930051X

TOSNE, LIZETETH. et al. Effect of Cassava Starch and Beeswax Coating on Chontaduro. Biotechnology in the Agricultural Sector and Hortic Agroindustry [On line]. 2014, (Colombia), 12 (2), pp. 6-7. Available at: http://www.scielo.org.co/pdf/bsaa/v12n2/v12n2a04.pdf

TRUJILLO RIVERA, Cinthya Tatiana. "Obtaining biodegradable films from cassava starch (manihot esculenta crantz) doubly modified for use in food packaging"[On line] (Engineering) (Degree Thesis). Universidad Nacional Amazónica De Madre De Dios, Facultad De Ingeniería, Escuela Académica Profesional De Ingeniería Agroindustrial. Puerto Maldonado - Peru. 2014. p.20. [Accessed: 2020-08-06]. Available at: http://repositorio.unamad.edu.pe/bitstream/handle/UNAMAD/65/004-2-1-013.pdf?sequence=1&isAllowed=y

VALENCIA CHAMORRO, S.; & TORRES MORALES, J. Edible coatings applied to IV and V gamma products. Iberoamerican Journal of Post-harvest Technology. [Online]. 2016, (Mexico), 17 (2), pp. 162-163-164-165. [Accessed: 8 July 2020]. ISSN 1665-0204. Available at: https://www.redalyc.org/jatsRepo/813/81349041004/html/index.html

VILLADA, Héctor. et al. Research on Thermoplastic Starches, Precursors of Biodegradable Products. Información Tecnológica. [Online], 2018, (Colombia), 19 (2), p.6. [Accessed: 03 August 2020]. Available at: https://scielo.conicyt.cl/pdf/infotec/v19n2/art02.pdf

ZAPADOR, M & CHIRAL, A. Starch-Based Coatings for Preservation of Fruits and Vegetables. Revestim [Online]. 2018, (Spain), 8 (5). pp.2-7. [Accessed: 10 July 2020]. DOI: 10.3390 / revestim 8050152. Retrieved from: https://translate.google.com/translate?hl=es&sl=en&u=https://www.researchgate.net/publication/324753468_Starch-Based_Coatings_for_Preservation_of_Fruits_and_Vegetables&prev=search&pto=aue

ZAPATA CRIOLLO Danixa Marilyn. Evaluation of biofilms formulated from green banana (Musa paradisiaca) and cassava (Manihot esculenta) starch with aloe vera gel [On line] (Engineering) (Degree Thesis). Universidad Nacional De Piura, Faculty of Industrial Engineering, Professional School of Agroindustrial and Food Industries. Piura-Peru. 2019. p. 16. [Accessed: 2020-08-03]. Available at: http://repositorio.unp.edu.pe/bitstream/handle/UNP/1586/IND-ZAP-CRI-2019.pdf?sequence=1&isAllowed=y

More
Books!

Printed by Books on Demand GmbH, Norderstedt / Germany